Collins

AQA GCSE 9-1
Maths Higher

Practice Papers

Mike Fawcett and Keith Gordon

Contents

SET A

SET B

ANSWERS

Acknowledgements

The authors and publisher are grateful to the copyright holders for permission to use quoted materials and images.

All images are © HarperCollins*Publishers* and Shutterstock.com

Every effort has been made to trace copyright holders and obtain their permission for the use of copyright material. The authors and publisher will gladly receive information enabling them to rectify any error or omission in subsequent editions. All facts are correct at time of going to press.

Published by Collins
An imprint of HarperCollins*Publishers*
1 London Bridge Street
London SE1 9GF

HarperCollins*Publishers*
Macken House, 39/40 Mayor Street Upper,
Dublin 1, D01 C9W8

Commissioning Editor: Kerry Ferguson
Project Leaders: Chantal Addy and Richard Toms
Authors: Mike Fawcett and Keith Gordon
Consultant Author: Trevor Senior
Cover Design: Sarah Duxbury and Kevin Robbins
Inside Concept Design: Ian Wrigley
Text Design and Layout: QBS Learning
Production: Karen Nulty
Printed in the United Kingdom

MIX
Paper
FSC™ C007454

© HarperCollins*Publishers* 2024

Collins

AQA
GCSE
Mathematics

H

SET A – Paper 1 Higher Tier

Author: Mike Fawcett

Materials

Time allowed: 1 hour 30 minutes

For this paper you must have:

- mathematical instruments

You may **not** use a calculator.

Instructions

- Use black ink or black ball-point pen. Draw diagrams in pencil.
- Answer **all** questions.
- You must answer the questions in the space provided.
- In all calculations, show clearly how you work out your answer.

Information

- The marks for questions are shown in brackets.
- The maximum mark for this paper is 80.
- You may use additional paper, graph paper and tracing paper.

Name: ....

Answer **all** questions in the spaces provided.

1 Volume of a cone = $\frac{1}{3}\pi r^2 h$

Work out the volume of a cone with base radius 6 cm and height 8 cm.
Give your answer in terms of π.

[2 marks]

Answer _____ cm³

2 Expand $4x^3(2x + y)$

[2 marks]

Answer _____

3 Solve the inequality $5x + 3 < -2$

[2 marks]

Answer _____

4 A snooker cue is $58\frac{3}{4}$ inches long.

A piece is cut off the bottom and it now measures $37\frac{2}{5}$ inches.

Work out the length of the piece that is cut off.

[3 marks]

Answer .... inches

5 Write 54 as a product of its prime factors.
Give your answer in index form.

[3 marks]

Answer ....

6   Tennis balls cost 48p each or packs of 3 for £1.25

What is the smallest possible cost for 40 tennis balls?

[4 marks]

Answer £ ....

7   Four identical cuboids measuring 3 ft × 3 ft × 8 ft are pushed together to form a rectangular shape with a hollow centre.

The plan of the shape is shown.

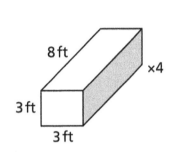

Plan

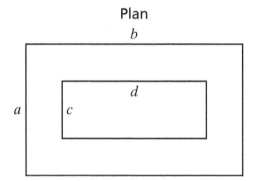

Write down the lengths of a, b, c and d.

[4 marks]

$a =$ .... ft

$b =$ .... ft

$c =$ .... ft

$d =$ .... ft

8 Rachel is at the gym for two hours.

She spends $\frac{2}{5}$ of her time on the weights.

The rest of her time is spent running and cycling in the ratio of 4 : 5

How many minutes does she spend cycling?

[4 marks]

Answer _____ minutes

9 Shape *ABCD* is a trapezium.

CDE and *GDF* are both straight lines.

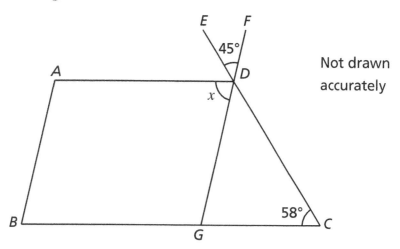

Not drawn accurately

Work out the size of angle *x*.

Give reasons for any angles you write down or calculate.

[3 marks]

x = _____ degrees

10 A machine can make 53 items in 5.8 minutes.

10 **(a)** Estimate the number of items the machine can make in one day.

[2 marks]

Answer

10 **(b)** State any assumptions that you have made.

[1 mark]

11 **(a)** Given the formula $v^2 = u^2 + 2as$

Work out the value of v when $u = 4$, $a = 3$ and $s = -2$

[3 marks]

$v = $

11 **(b)** Make a the subject of the formula $v^2 = u^2 + 2as$

[2 marks]

12 Look at the following vector statement.

$$\begin{pmatrix} 4 \\ 3 \end{pmatrix} - \begin{pmatrix} a \\ 2b \end{pmatrix} = \begin{pmatrix} 6 \\ 2 \end{pmatrix}$$

Work out the values of a and b.

[2 marks]

$a = $ ....

$b = $ ....

13 Work out the value of $8^{\frac{2}{3}}$

[2 marks]

Answer ....

14 In a game, Ethan, Bob and Josh throw a bottle and try to land it upright.

Here are the results.

	Ethan	Bob	Josh
Number of tries	10	25	50
Number of lands upright	1	3	4

14 **(a)** Who is the best at the game?
Give a reason to support your answer.

[2 marks]

14 **(b)** Whose results give you the best understanding of their ability?
Give a reason for your decision.

[2 marks]

15 Work out the nth term for the following sequence.

$-1 \qquad 5 \qquad 15 \qquad 29 \qquad 47 \qquad ...$

[2 marks]

Answer

16   Ali, Brad and Dora each have some marbles.

Brad has 4 times as many as Ali. Dora has 12 more than Ali.

Together, Ali and Brad have the same number of marbles as Dora.

How many marbles does Ali have?

[3 marks]

Answer _____

17   A group of students are asked whether or not they walk to school.

The table shows some of the information.

	Walks	Does not walk	Total
Boys	7		13
Girls			
Total	11	15	

17   (a) Complete the table.

[3 marks]

17   (b) A girl is chosen at random.

What is the probability that she does **not** walk to school?

[1 mark]

Answer _____

18 Tim cycles along a road to test his new bike.

He stops on the way to adjust his brakes.

The graph shows his journey along the road.

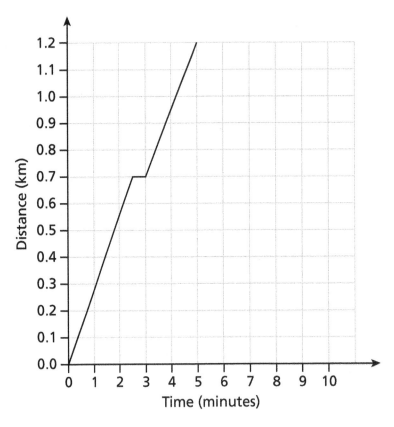

18 **(a)** How long did Tim spend adjusting his brakes?

[1 mark]

Answer ....

18 **(b)** Work out his average speed from home to the end of the road.
Give your answer in metres per second (m/s).

[2 marks]

....

....

....

Answer .... m/s

18 **(c)** Tim rests for 1 minute at the end of the road before cycling back.

He then cycles back at 6 m/s.

Complete the graph for his journey.

[3 marks]

19 Expand and simplify $(x + 2)(x + 3)(x - 1)$

[4 marks]

Answer ....

20   The table shows the length of songs on the top five albums in the UK charts.

Length, l (mins)	$0 < l \leq 1$	$1 < l \leq 2$	$2 < l \leq 3$	$3 < l \leq 4$	$4 < l \leq 5$	$5 < l \leq 6$	$6 < l \leq 7$
Frequency	3	5	7	18	17	7	3

20   **(a)**   Show this information on the cumulative frequency graph.

[3 marks]

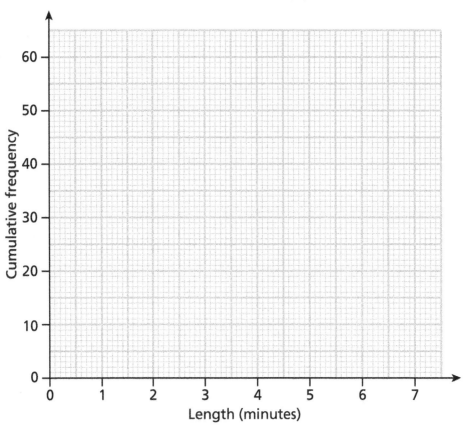

20   **(b)**   Use the graph to estimate the median length of song.

[1 mark]

Answer _____ minutes

20   **(c)**   Draw a box plot to illustrate the data.

[2 marks]

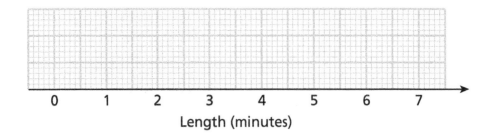

21

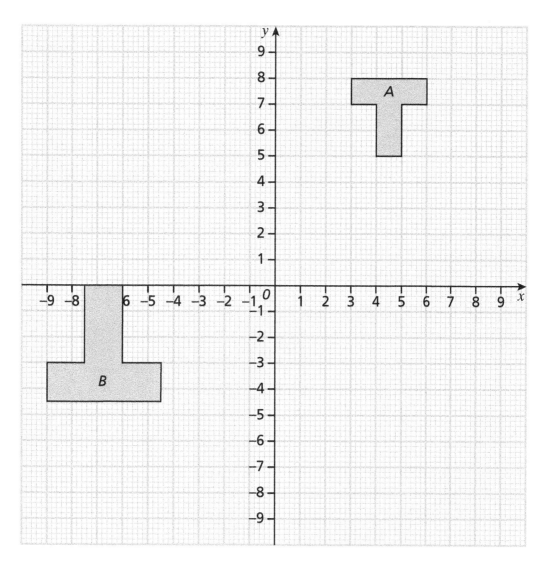

Describe the single transformation which maps shape *A* onto shape *B*.

[3 marks]

22 Two measuring jugs are mathematically similar.

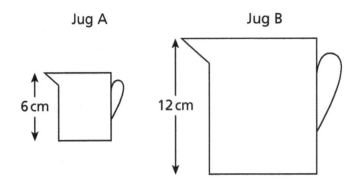

Jug A holds 300 ml.

How much will Jug B hold?
Give your answer in litres.

[3 marks]

Answer .... litres

23 The diagram shows a circle with centre (0, 0).

A tangent *AB* touches the circle at the point *P* (7, 10).

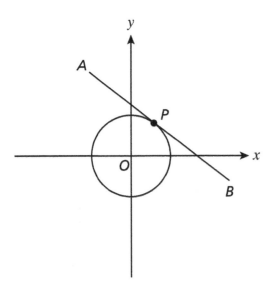

23 **(a)** Write down the gradient of *OP*.

[1 mark]

Answer _____

23 **(b)** Write down the gradient of *AB*.

[1 mark]

Answer _____

23 **(c)** Work out the equation of the line *AB*.

[2 marks]

_____

_____

Answer _____

24 Show that the perimeter of the triangle can be written in the form $a\sqrt{b} + c$

[2 marks]

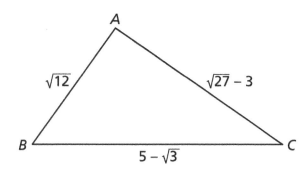

_____

_____

_____

END OF QUESTIONS

THIS PAGE HAS DELIBERATELY BEEN LEFT BLANK

Collins

AQA
GCSE

Mathematics

H

SET A – Paper 2 Higher Tier

Author: Mike Fawcett

Materials

Time allowed: 1 hour 30 minutes

For this paper you must have:

- calculator
- mathematical instruments

Instructions

- Use black ink or black ball-point pen. Draw diagrams in pencil.
- Answer **all** questions.
- You must answer the questions in the space provided.
- In all calculations, show clearly how you work out your answer.

Information

- The marks for questions are shown in brackets.
- The maximum mark for this paper is 80.
- You may use additional paper, graph paper and tracing paper.

Name: _____

Answer **all** questions in the spaces provided.

1 Write 350 ml to 1.2 litres as a ratio in its simplest form.

[2 marks]

Answer _____

2 Work out the area of the semicircle.
Give your answer as a multiple of π.

[2 marks]

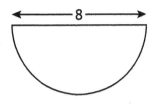

Answer _____

3 Simplify fully $\dfrac{x^5 \times x^{-1}}{x^2}$

[2 marks]

Answer _____

4 (a) Doug is going to measure the height of some students in order to analyse any differences between boys and girls.

Which of the following statements best describe Doug's data?

Tick **two** boxes.

[1 mark]

Primary ☐

Secondary ☐

Discrete ☐

Continuous ☐

4 (b) Doug is going to select the students from his class at random.

Describe how he could do this.

[1 mark]

_____

_____

5 Show that the area of the triangle can be written as $x^2 + 2x - 8$

[2 marks]

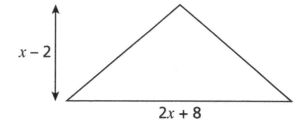

_____

_____

_____

_____

6 Louise and Anita work in a café.

Louise takes 42 seconds to make a latte.

Anita takes 70 seconds to toast a teacake.

Louise says, 'I can make x lattes in the time it takes you to toast y teacakes.'

Work out the values of x and y.

[3 marks]

$x =$ ....

$y =$ ....

7 Translate shape A with the vector $\begin{pmatrix} -2 \\ -4 \end{pmatrix}$ and label the new shape B.

[2 marks]

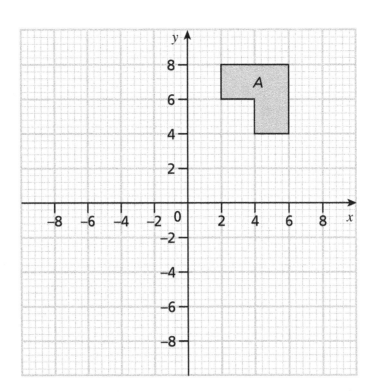

8   (a) Emma buys a house for £202 000.

The house increases in value at 1.5% per annum.

How many years will it be until the value of the house exceeds £215 000?

[2 marks]

Answer _____ years

8   (b) Cathy wants to build a new driveway and have a loft conversion on her house.

She is told that:

a new driveway will add 6% on to the value of her house

a loft conversion will add **a further** 18% on to the value of her house.

Cathy says, 'My house will be worth £180 000 if I have all of the work done'.

What is the current value of her house?
Give your answer to 3 significant figures.

[3 marks]

Answer £ _____

9   **(a)**   Given that $3.2 \times 10^7 \times A = 2.176 \times 10^4$

Write down the value of A as an ordinary number.

[3 marks]

$A =$ ....

9   **(b)**   A spider exerts a downward force of $1.15 \times 10^{-3}\,\text{N}$

Each of its feet has an area of $2.3 \times 10^{-5}\,\text{m}^2$

$$\boxed{Force = Pressure \times Area}$$

Work out the pressure applied to each of the spider's feet.

[2 marks]

Answer = ... N/m^2

10 There are 20 students in Anand's class.

He copies down this table from the board to show the heights of everyone in the class.

He has made one error in the frequency column.

Height, h (cm)	Frequency
$140 < h \leq 150$	3
$150 < h \leq 160$	6
$160 < h \leq 170$	6
$170 < h \leq 180$	4

The teacher says, 'An estimate for the mean height is 161 cm.'

Which class interval has an incorrect frequency?
You must show working to back up your answer.

[4 marks]

Answer ....

11 Solve $x^2 + 2x - 15 = 0$

[2 marks]

$x =$ ....

$x =$ ....

12 **(a)** By plotting the graph of $3y = 5x + 3$, solve the simultaneous equations

$$3y = 5x + 3$$

$$x + y = 5$$

[4 marks]

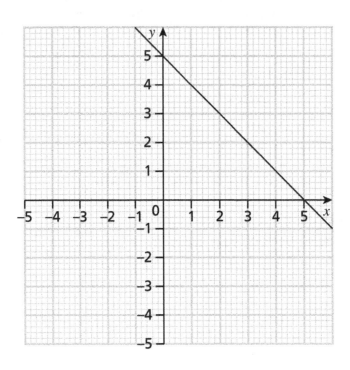

$x =$ ....

$y =$ ....

12 **(b)** Work out the equation of the line which is parallel to $x + y = 5$ and goes through the point (3, 4).

[2 marks]

....

....

....

Answer ....

13   Each time you pot a ball in snooker, you get to have another shot.

The probability that Craig pots a ball is 0.23

13   **(a)**   Work out the probability that Craig has exactly three shots on his next turn.

[2 marks]

Answer

13   **(b)**   The probability that Ed will pot two balls in a row is 0.0961

What is the probability that Ed will miss any given ball that he goes for?

[2 marks]

Answer

14 The area of this sector of a circle is 2.5π cm².

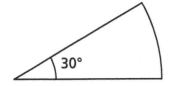

Work out the radius of the circle.
Give your answer to 2 decimal places.

[3 marks]

Answer _____ cm

15 The graph shows the height of water in a container which is left out overnight in the rain.

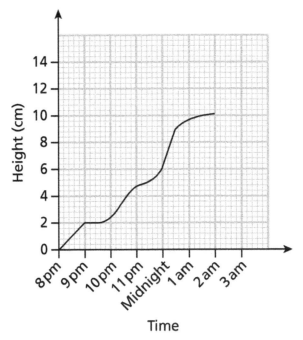

15 (a) At what time was the rainfall the heaviest?

[1 mark]

Answer _____

15 **(b)** Estimate the rate of rainfall at 10:30 pm.

[2 marks]

Answer _____ cm/hour

16 y is directly proportional to the cube root of x.

16 **(a)** Use the table to find an equation for y in terms of x.

[2 marks]

x	0	8	64
y	0	5	10

Answer _____

16 **(b)** Calculate the value of x when $y = 15$.

[2 marks]

$x =$ _____

17 A restaurant claims to have 455 different combinations when you buy a three-course meal.

The restaurant serves five different starters.

What is the total number of mains and desserts that the restaurant serves?

[2 marks]

Answer ....

18 **(a)** Majid completes a 400 m sprint in 50 seconds.

The velocity–time graph shows his run.

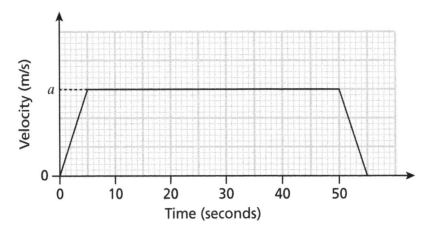

Calculate the value of a.
Give your answer to 2 decimal places.

[2 marks]

$a =$ .... m/s

18 (b) The histogram shows the time that it took all of the runners to complete the 400 m sprint.

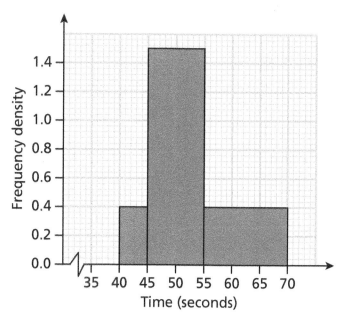

How many runners took part in the competition?

[3 marks]

Answer ....

19 (a) Work out the turning point of the graph of $y = 4x^2 - 5x + 12$

[4 marks]

Answer (...................... , ...)

19 **(b)** The graph of $y = f(x)$ is shown with a turning point of (2, 3).

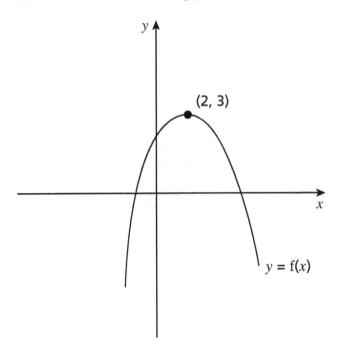

Work out the turning point of the graph of $y = f(x - 3)$

[2 marks]

Answer (........................ , ...)

20 An Olympic pool is 50 m long to the nearest centimetre.

Jenny can swim four lengths in 2 minutes and 15 seconds to the nearest second.

By considering bounds, give Jenny's speed to a suitable degree of accuracy.

[4 marks]

Answer .... m/s

21 Prove that the product of three consecutive even numbers is always divisible by 8.

[3 marks]

22 *ABCD* is an isosceles trapezium.

Point *E* is on a straight line with *BC* such that *ABED* is a trapezium containing two right angles.

$\overrightarrow{AB} = \mathbf{a}$

$\overrightarrow{AD} = \mathbf{b}$

$BC : AD = 3 : 4$

Write the vector $\overrightarrow{AE}$ in terms of **a** and **b**.

[4 marks]

Answer

23 Shape *ACD* is a triangle.

AC = 10.8 cm

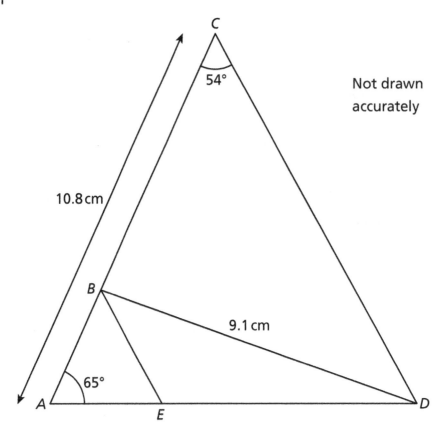

Not drawn accurately

Calculate the area of triangle *BCD*.
Give your answer to 1 decimal place.   **[5 marks]**

Answer .... cm²

END OF QUESTIONS

Collins

AQA
GCSE
Mathematics

H

SET A – Paper 3 Higher Tier

Author: Mike Fawcett

Materials

Time allowed: 1 hour 30 minutes

For this paper you must have:

- calculator
- mathematical instruments

Instructions

- Use black ink or black ball-point pen. Draw diagrams in pencil.
- Answer **all** questions.
- You must answer the questions in the space provided.
- In all calculations, show clearly how you work out your answer.

Information

- The marks for questions are shown in brackets.
- The maximum mark for this paper is 80.
- You may use additional paper, graph paper and tracing paper.

Name: ....

Answer **all** questions in the spaces provided.

1   Decrease £350 by 13%

[2 marks]

Answer

2   Write down the name of this type of sequence.

[1 mark]

6   12   24   48   96

Answer

3   Work out the gradient of the line $2x - y + 8 = 0$

[2 marks]

Answer

4 Work out the three-figure bearing of *A* from *B*.

[2 marks]

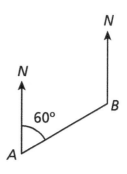

Answer .... °

5 The diagram shows two similar rectangles.

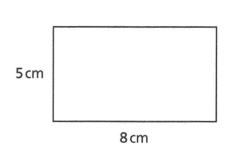

7.5 cm

5 cm

8 cm

Work out the length of the larger rectangle.

[2 marks]

Answer .... cm

6   10 people were asked their height and annual income.

The tables show the results.

Income (£)	14 000	21 000	26 500	32 500	28 500
Height (m)	1.59	1.72	1.85	1.65	1.57

Income (£)	15 000	13 000	25 000	33 500	29 000
Height (m)	1.83	1.71	1.65	1.79	1.72

6   **(a)** Plot a scatter graph for this data.

[2 marks]

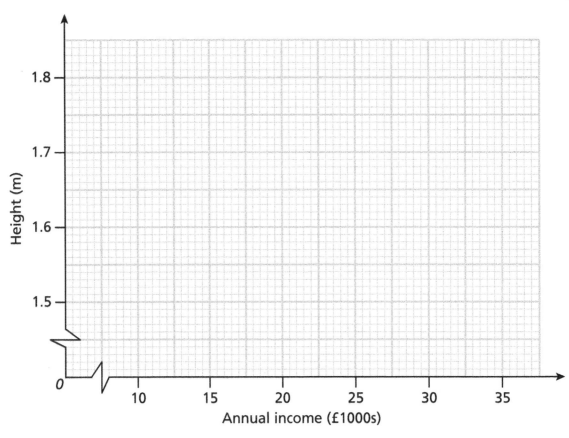

6   **(b)** Interpret the scatter graph, stating any correlation found.

[2 marks]

7   Work out the highest common factor (HCF) of 135 and 630.

[3 marks]

Answer

8   A sports centre has a gym and a swimming pool.

On Wednesday, 51 people visited the centre.

13 people who used the gym also went swimming.

21 people did not use the gym and 6 of those did not swim either.

Complete the frequency tree to illustrate this information.

[2 marks]

9   26 people start on an evening course in September.

By January only 19 people remain on the course.

What percentage of people have left the course?

[2 marks]

Answer _____ %

10 A metal rod for a piece of machinery is $3\frac{4}{5}$ inches long.

The designer says that the rod would work better if it was $\frac{1}{3}$ longer.

How long should the rod be?
Give your answer as a mixed number.

[3 marks]

Answer _____ inches

11 The interior angle of a regular polygon is $2x$.

Show that the number of sides that the polygon has can be written in the form $\dfrac{a}{b-x}$

[3 marks]

12 Ed prints metallic signs in three sizes.

The tables show the prices he charges and his costs for printing and postage.

Size	Price
A5	£5.95
A4	£8.65
A3	£10.85

Size	Printing costs
A5	£1.07
A4	£1.52
A3	£3.09

Number of items per customer	Postal costs
1 sign	£2.40
2 or more signs	£3.80

Ed has this offer

- 3 for the price of 2
- free delivery

He makes these sales to three customers.

- Two A3 signs and one A4 sign
- Two A5 signs
- One A4 sign

How much profit did Ed make from the three customers?

[6 marks]

Answer £ ....

13 An airfield is going to be built near the towns of Brooks, Redding and Dufresne.

It must be closer to Redding than to Brooks.

It must be within 5 miles of Dufresne.

Scale:
1 cm = 1.6 miles

Redding
✗

✗
Dufresne

✗
Brooks

Shade on the map, the region where the airfield can be built.

[3 marks]

14 Solve the inequality $2x^2 - 5x \leqslant 3$

[4 marks]

Answer

15 11 people are asked how many songs they downloaded this month.

Here are the results:

15, 21, 18, 17, 15, 17, 20, 5, 13, 6, 26

Jason says, 'The median is closer to the upper quartile than to the lower quartile.'

Is he correct?
You **must** show your working.

[4 marks]

Answer ....

16 2.5 litres of water is boiling in a pan, with a constant heat.

The water evaporates at a rate of 3.5% every minute.

How many minutes would it take for 500 ml of water to evaporate?

[3 marks]

Answer .... minutes

17 **(a)** Which graph shows the relationship that y is inversely proportional to x?

Circle your answer.

[1 mark]

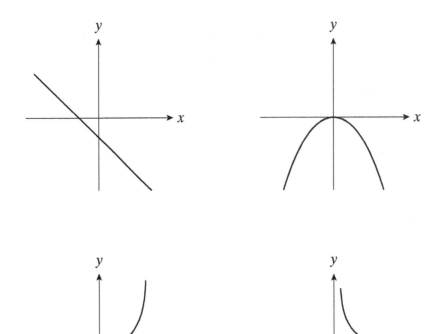

17 **(b)** Sketch a graph to show that y is directly proportional to the square of x.

[1 mark]

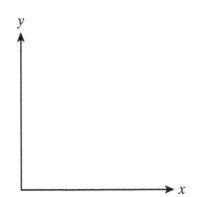

18 *A, B, C* and *D* are points on the circumference of a circle, with centre *O*.

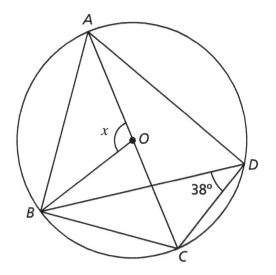

Not drawn accurately

Work out the size of the angle marked *x*.
Give reasons for your answer.

[4 marks]

_____

_____

_____

_____

_____

x = _____ degrees

19 Prove that $0.2\dot{3}$ can be written as $\dfrac{7}{30}$

[3 marks]

_____

_____

_____

_____

20 **(a)** Show that the equation $x^3 + 5x - 3 = 0$ can be rearranged into $x = \dfrac{3 - x^3}{5}$

[2 marks]

20 **(b)** Show that the equation $x^3 + 5x - 3 = 0$ has a solution between 0 and 1.

[2 marks]

20 **(c)** Starting with $x_0 = 0$

Use the iteration formula $x_{n+1} = \dfrac{3 - x_n^3}{5}$ to work out a solution to the equation $x^3 + 5x - 3 = 0$

Give your answer correct to 2 decimal places.

[3 marks]

$x = $ ....

21   The universal set = {factors of 100}

Set A = {prime numbers}

Set B = {even numbers}

21   **(a)** Complete the Venn diagram.

[3 marks]

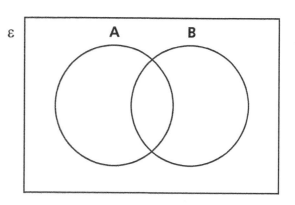

21   **(b)** Work out P(A ∩ B)

[1 mark]

Answer

22   The first four terms of a geometric sequence are

$\sqrt{2}$ 2 $2\sqrt{2}$ 4

Find the 9th term in this sequence.
Give your answer as a surd in its simplest form.

[2 marks]

Answer

23 Solve the simultaneous equations.

$y = x + 1$

$y = x^2 - 5x + 10$

[4 marks]

$x =$ _____ $y =$ _____

24 Simplify fully $\dfrac{5x^2 - 20}{x^2 - 2x}$

[3 marks]

Answer _____

25 The diagram shows a cube.

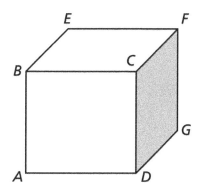

25 **(a)** Using Pythagoras' theorem, work out the ratio *AD* : *AF*.

[3 marks]

Answer .... : ....

25 **(b)** Work out the size of angle *GAF*.

[2 marks]

GAF = ... degrees

END OF QUESTIONS

THIS PAGE HAS DELIBERATELY BEEN LEFT BLANK

Collins

AQA

GCSE

Mathematics

H

SET B – Paper 1 Higher Tier

Author: Keith Gordon

Materials

Time allowed: 1 hour 30 minutes

For this paper you must have:

- mathematical instruments

You may **not** use a calculator.

Instructions

- Use black ink or black ball-point pen. Draw diagrams in pencil.
- Answer **all** questions.
- You must answer the questions in the space provided.
- In all calculations, show clearly how you work out your answer.

Information

- The marks for questions are shown in brackets.
- The maximum mark for this paper is 80.
- You may use additional paper, graph paper and tracing paper.

Name: ....

Answer **all** questions in the spaces provided.

1   f(x) = x − 3

1   **(a)** Write down the value of f(−0.5)

[1 mark]

Answer ....

1   **(b)** Write down an expression for f⁻¹(x)

[1 mark]

Answer ....

2   Write down the roots of the equation (x − 2)(x + 3) = 0

[1 mark]

Answer ....

3   **(a)** Here is a right-angled triangle *ABC*.

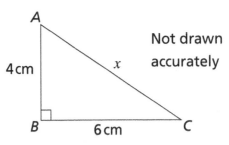

Work out the **exact** value of the length x.

[2 marks]

....

....

Answer ....cm

3 **(b)** Here is a right-angled triangle *PQR*.

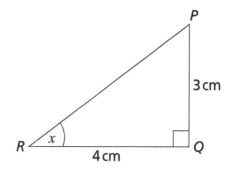

Not drawn accurately

Circle the value of the tangent of angle x.

[1 mark]

$\dfrac{3}{5}$ $\qquad\qquad$ $\dfrac{3}{4}$ $\qquad\qquad$ $\dfrac{4}{5}$ $\qquad\qquad$ $\dfrac{4}{3}$

4 Solve $\quad 3(x-2)+4=\dfrac{x}{2}$

[3 marks]

$x =$ ....

5 Work out the surface area of this cuboid.

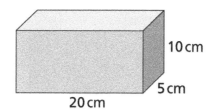

10 cm

5 cm

20 cm

[3 marks]

Answer .... cm²

6   Expand and simplify $4(x + 1) - 2(3x - 4)$

[3 marks]

Answer ....

7   **(a)**   Write 2.3×10^5 as an ordinary number.

[1 mark]

Answer ....

7   **(b)**   Write 0.0005 in standard form.

[1 mark]

Answer ....

7   **(c)**   Work out $(2 \times 10^4) \times (8 \times 10^3)$

Give your answer in standard form.

[2 marks]

Answer ....

8 The graph of $y = 2x^2 - 3x - 5$ is shown.

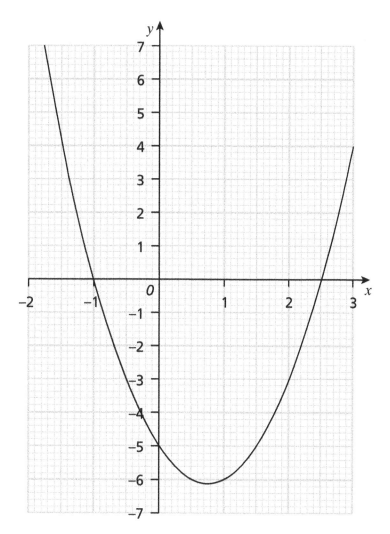

8 **(a)** Write down the values of x when $y = 4$.

[2 marks]

Answer .... and ....

8 **(b)** Write down the coordinates of the turning point.

[1 mark]

Answer (................................ , ...)

9   Part of a regular polygon is shown.

How many sides does the polygon have?

[3 marks]

Answer ....

10   Here is a square.

$(x + 2)\,\text{cm}$

Not drawn accurately

$(2x - 1)\,\text{cm}$

Work out the area.
You **must** show your working.

[4 marks]

Answer .... cm^2

11 Expand $(x + 1)^2(x - 3)$

[3 marks]

Answer

12 The diagram shows a cylinder with diameter of base $2x$ cm and height 3 cm.

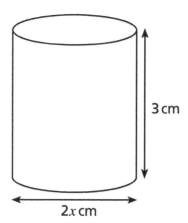

3 cm

$2x$ cm

Its volume is 48π cm³.

Work out the **radius** of the base.

[3 marks]

Answer _____ cm

13 Write down the three inequalities that define the region *R*.

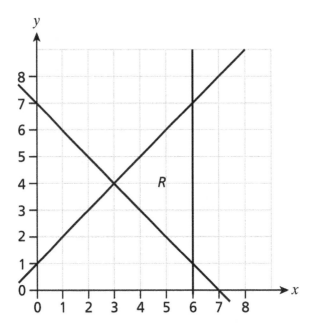

[3 marks]

Answer ....

....

....

14 Expand and simplify $(3+\sqrt{2})(9-\sqrt{8})$

Give your answer in the form $a+b\sqrt{2}$, where a and b are integers.

[3 marks]

Answer ....

15 Draw a histogram for the data below.

Height, h (cm)	Frequency
$5 \leqslant h < 10$	15
$10 \leqslant h < 20$	35
$20 \leqslant h < 35$	30
$35 \leqslant h < 45$	15
$45 \leqslant h < 50$	5

[3 marks]

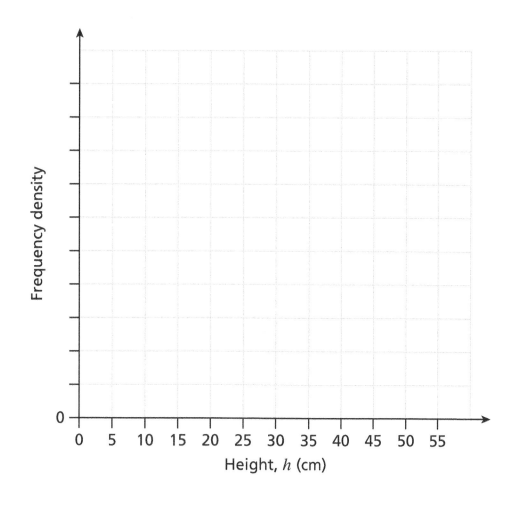

16 **(a)** *O* is the centre of the circle.

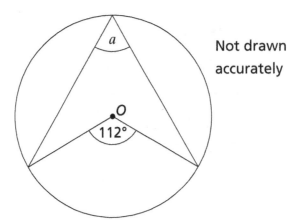

Not drawn accurately

Work out the size of angle *a*.

[1 mark]

Answer .... degrees

16 **(b)** *ABC* are points on the circumference of a circle, centre *O*.

SAT is a tangent.

BC is a diameter.

Angle *BAT* = 32°

Not drawn accurately

Work out the size of angle *CBA*, marked x on the diagram.
You **must** show your working, which may be on the diagram.

[3 marks]

Answer _____ degrees

17 Work out $64^{\frac{2}{3}}$

[2 marks]

Answer _____

18 The cumulative frequency diagram shows the ages of people at a wedding.

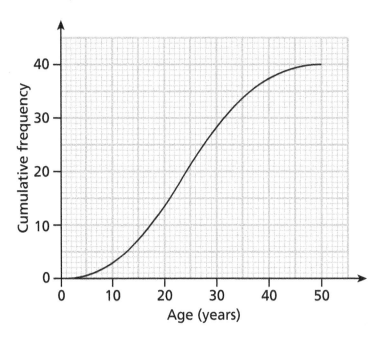

18 **(a)** Write down an estimate of the median age.

[1 mark]

Answer _____ years

18 **(b)** Work out an estimate of the interquartile range.

[2 marks]

Answer _____ years

18 **(c)** The youngest person at the wedding was 5 years old.

Draw a box plot for the data.

[2 marks]

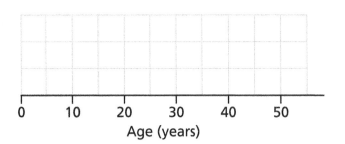

19 $\overrightarrow{OA} = \mathbf{a}$

$\overrightarrow{AB} = \dfrac{3}{2}\,\mathbf{b}$

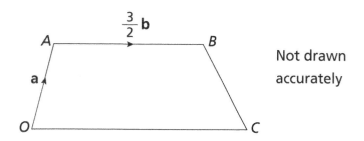

Not drawn
accurately

19   **(a)**   Write down the vector $\overrightarrow{OB}$ in terms of **a** and **b**.

[1 mark]

Answer ....

19   **(b)** $\overrightarrow{BC} = -\mathbf{a} + \dfrac{1}{2}\mathbf{b}$

Work out the vector $\overrightarrow{OC}$.

[2 marks]

Answer ....

19   **(c)**   Is *OABC* a trapezium?
Give a reason for your answer.

[1 mark]

20 Write the recurring decimal 3.733333…. as a mixed number.

Answer

21 The area of a right-angled isosceles triangle is 9 cm².

Not drawn
accurately

Work out the perimeter of the triangle.
Give your answer in the form $a + b\sqrt{c}$, where a, b and c are integers.

[5 marks]

Answer _____ cm

22 A bag contains 10 counters.

7 of the counters are red, 3 of them are blue.

Two counters are taken from the bag.

Work out the probability that they are different colours.

[5 marks]

Answer ....

23 Simplify fully $\dfrac{4x^2 - 4x - 15}{4x^2 - 9}$

[3 marks]

Answer ....

24 $A(3, 10)$ and $B(7, 8)$ are two points.

Work out the equation of the line that is

perpendicular to AB

passes through the midpoint of AB.

[5 marks]

Answer ....

END OF QUESTIONS

Collins

AQA

GCSE

Mathematics

SET B – Paper 2 Higher Tier

Author: Keith Gordon

Materials

Time allowed: 1 hour 30 minutes

For this paper you must have:

- calculator
- mathematical instruments

Instructions

- Use black ink or black ball-point pen. Draw diagrams in pencil.
- Answer **all** questions.
- You must answer the questions in the space provided.
- In all calculations, show clearly how you work out your answer.

Information

- The marks for questions are shown in brackets.
- The maximum mark for this paper is 80.
- You may use additional paper, graph paper and tracing paper.

Name: _____

Answer **all** questions in the spaces provided.

1 Work out $\frac{3}{4} \times 12$

[1 mark]

Answer

2 Write down a fraction that is a recurring decimal.

[1 mark]

Answer

3 The point $A(6, 7)$ is reflected in the y-axis to the point A'.

Write down the coordinates of A'.

[1 mark]

Answer (_____ , _____)

4

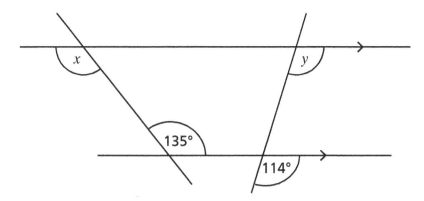

4 **(a)** Write down, with a reason, the size of angle x.

[1 mark]

$x =$ .... degrees

Reason ....

4 **(b)** Write down, with a reason, the size of angle y.

[1 mark]

$y =$ .... degrees

Reason ....

5 Translate the triangle by $\begin{pmatrix} -3 \\ -4 \end{pmatrix}$

[2 marks]

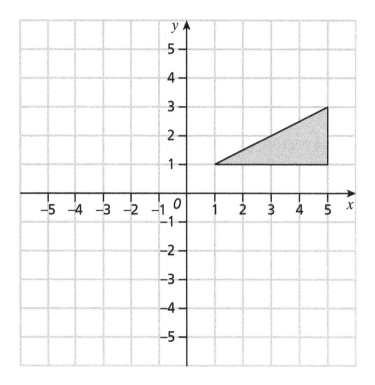

6 Work out the length x in the triangle.

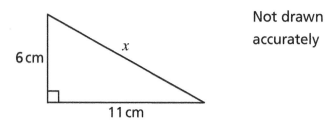

Not drawn accurately

[3 marks]

....

....

....

....

$x =$ .... cm

7 The table shows the heights of some young trees.

Height, h (cm)	Frequency
$140 \leqslant h < 150$	5
$150 \leqslant h < 160$	9
$160 \leqslant h < 170$	12
$170 \leqslant h < 180$	8
$180 \leqslant h < 190$	6

Work out an estimate of the mean height.

[3 marks]

Answer _____ cm

8 **(a)** As a product of prime factors $20 = 2^2 \times 5$

Work out 28 as a product of prime factors.

[2 marks]

Answer _____

8 **(b)** Work out the least common multiple of 20 and 28.

[2 marks]

Answer _____

9 Triangles *ABC* and *PQR* are similar.

Work out the value of x.

[4 marks]

$x =$ _____

10 A washing machine is reduced by 15% in a sale.

The sale price of the washing machine is £238.

What was the original price of the washing machine?

[3 marks]

Answer £ _____

11 Two numbers are in the ratio 2 : 5

The difference between the numbers is 36.

Work out the values of the two numbers.

[3 marks]

Answer _____ and _____

12 The area of this semicircle is 201 cm² to 3 significant figures.

Not drawn
accurately

Work out the perimeter of the semicircle.

[3 marks]

Answer _____ cm

13 Using ruler and compasses only, construct an angle of 30° at A.

You **must** show your construction arcs.

[3 marks]

A _____

14 **(a)** Expand and simplify $5(x - 2)(4x + 3)$

[3 marks]

Answer _____

14 **(b)** Factorise fully $2x^2 + 8x + 6$

[2 marks]

Answer _____

15   Enlarge the triangle by a scale factor of $-\dfrac{1}{3}$ about the centre (5, 8).

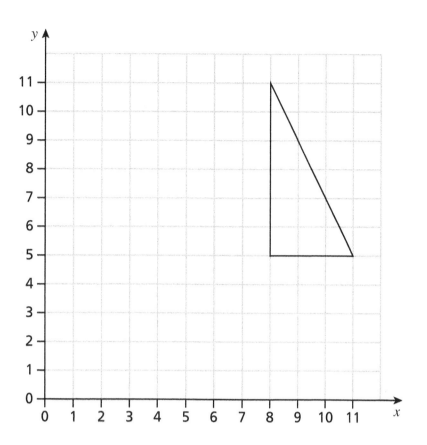

16 A jar contains 30 red beads and 40 white beads.

The number of red beads is increased by 60%

The number of white beads is increased by $p\%$

The number of red and white beads is now equal.

Work out the value of p.

[3 marks]

Answer ....

17 **(a)** Write $x^2 + 6x - 9$ in the form $(x + a)^2 - b$, where a and b are integers.

[3 marks]

Answer ....

17 **(b)** Hence, or otherwise, solve $x^2 + 6x - 9 = 0$

Give answers in the form $p \pm \sqrt{q}$, where p and q are integers.

[2 marks]

Answer ....

18 Write the equation $\dfrac{2}{x+1} - \dfrac{3}{4x-1} = 1$

in the form $ax^2 + bx + c = 0$ where a, b and c are integers.

[4 marks]

Answer ....

19 y is directly proportional to the square of x.

When $y = 20$, $x = 2$

19 **(a)** Work out the value of y when $x = 10$

[3 marks]

Answer ....

19 **(b)** Work out the value of x when $y = 5$

[2 marks]

Answer ....

20 146 students in year 7 were asked if they had a cat, a dog or both.

The Venn diagram shows the results.

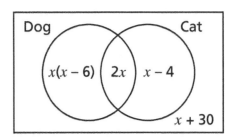

A student is picked at random.

Work out the probability that the student only has a cat.

[5 marks]

Answer ....

21   **(a)**   Work out angle x in this triangle.

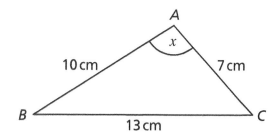

[3 marks]

....

....

....

$x =$ .... degrees

21   **(b)**   Work out the area of this triangle.

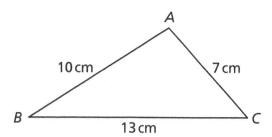

[2 marks]

....

....

....

Answer .... cm²

22 The speed–time graph for a journey is shown.

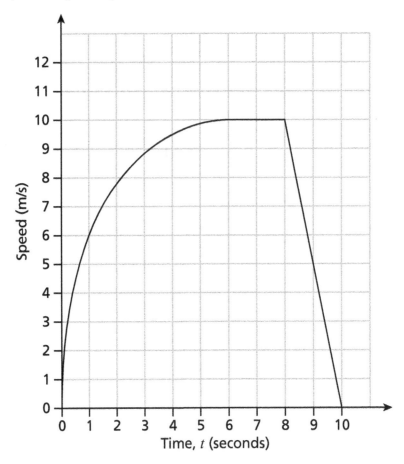

Time, t (seconds)

22 **(a)** Estimate the acceleration at 3 seconds.

[3 marks]

....

....

....

....

Answer .... m/s^2

22 (b) Estimate the average speed for the journey.

[4 marks]

Answer .... m/s

23 The formula connecting the sine of angle x, the opposite side (o) and the hypotenuse (h), is $\sin x = \dfrac{o}{h}$

$h = 12$ to 2 significant figures

$o = 8.3$ to 2 significant figures

Work out the upper and lower bounds for the angle x.
Give your angles to 1 decimal place.
You **must** show your working.

[5 marks]

Upper bound ....

Lower bound ....

END OF QUESTIONS

THIS PAGE HAS DELIBERATELY BEEN LEFT BLANK

Collins

AQA
GCSE
Mathematics

H

SET B – Paper 3 Higher Tier

Author: Keith Gordon

Materials

Time allowed: 1 hour 30 minutes

For this paper you must have:

- calculator
- mathematical instruments

Instructions

- Use black ink or black ball-point pen. Draw diagrams in pencil.
- Answer **all** questions.
- You must answer the questions in the space provided.
- In all calculations, show clearly how you work out your answer.

Information

- The marks for questions are shown in brackets.
- The maximum mark for this paper is 80.
- You may use additional paper, graph paper and tracing paper.

Name: _____

Answer **all** questions in the spaces provided.

1 Write down the value of 5^3

[1 mark]

Answer ....

2 Write 15 : 27 in the form 1 : n

[1 mark]

Answer .... : ....

3 Here is a sequence of consecutive cube numbers:

27   64   125   216

Write down the next cube number.

[1 mark]

Answer ....

4 Here are eight numbers.

  3   8   6   9   11   12   5   2

Work out the mean of the numbers.

[2 marks]

Answer _____

5 **(a)** Simplify $x^3 \times x^6$

[1 mark]

Answer _____

5 **(b)** Simplify $x^{12} \div x^2$

[1 mark]

Answer _____

6 Here are two column vectors:

$$a = \begin{pmatrix} 2 \\ 3 \end{pmatrix} \qquad b = \begin{pmatrix} 6 \\ -2 \end{pmatrix}$$

Work out $2a + b$.

[2 marks]

Answer _____

7

Write down the integers that are satisfied by the inequality shown.

[2 marks]

8 Enlarge the shape by a scale factor of $\frac{1}{3}$

[2 marks]

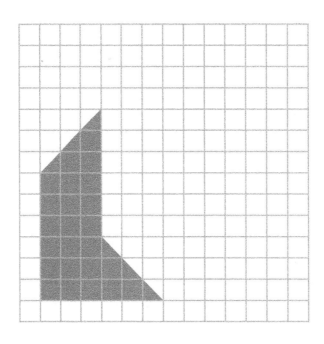

9 Solve the simultaneous equations

$3x + 2y = 2$

$x + 4y = 9$

[3 marks]

$x = $ ....

$y = $ ....

10 A bag contains 10 balls.

4 of the balls are red and 6 are blue.

A ball is chosen at random and then replaced.

Another ball is then chosen at random from the bag.

10 **(a)** Complete the tree diagram.

[1 mark]

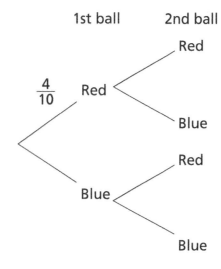

1st ball   2nd ball

$\frac{4}{10}$ Red Red

Blue

Blue Red

Blue

10 **(b)** Work out the probability that both balls were the same colour.

[3 marks]

Answer

11 A large candle exerts a pressure of 2 Pa on its base.

As the candle burns the pressure decreases.

After 2 hours the pressure is 0.5 Pa

Work out the rate of change of pressure.
Give your answer in Pa/hour.

[2 marks]

Answer .... Pa/hour

12 **(a)** Factorise $x^2 - 25$

[1 mark]

Answer ....

12 **(b)** Show that $(x + 2)^2 - (x + 1)^2 \equiv 2x + 3$ **[3 marks]**

13 **(a)** Show that the length x in the triangle below is 6.36 cm to 2 decimal places.

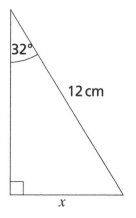

32°

12 cm

Not drawn
accurately

x

[2 marks]

13 **(b)** A cone has a half vertical angle of 32° and a slant height l of 12 cm.

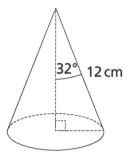

32° 12 cm

Work out the curved surface area of the cone.

The formula for the curved surface area of a cone is

Curved surface area = π × radius of base × slant height

[2 marks]

Answer .... cm²

14   A seal colony has 6000 seals.

The numbers are declining by 8% each year.

How many years will it be before the number of seals is below 3000?

[4 marks]

Answer .... years

15   Here are three graphs.

Graph A   Graph B   Graph C

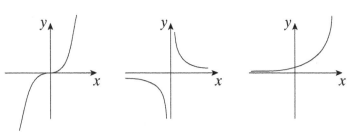

Match each graph to its possible equation.

[2 marks]

| Graph A |
| Graph B |
| Graph C |

| $y = \tan x$ |
| $y = x^2$ |
| $y = 2^x$ |
| $y = \dfrac{1}{x}$ |

16 Simplify $(2x^2y^3)^2$

[2 marks]

Answer

17 Here are the equations of four lines.

Line A: $y = 3x + 3$ Line B: $y = \dfrac{1}{4}x - 3$

Line C: $y = \dfrac{1}{3}x + 3$ Line D: $y = -4x - 4$

17 **(a)** Which two lines are perpendicular?

[1 mark]

Answer _____ and _____

17 **(b)** Which two lines intersect on the *x*-axis?

[1 mark]

Answer _____ and _____

18 **(a)** Write down the next two terms of this quadratic sequence.

[2 marks]

| 3 | 5 | 8 | 12 | 17 | 23 | ... | ... |

Answer _____ and _____

18 **(b)** Work out the nth term of the quadratic sequence.

| 6 | 10 | 16 | 24 | 34 | 46 | ... |

[4 marks]

Answer _____

19 The triangle *A*, shown, is reflected in *y* = 6

Call this triangle *B*.

Triangle *B* is then reflected in *x* = 5

Call this triangle *C*.

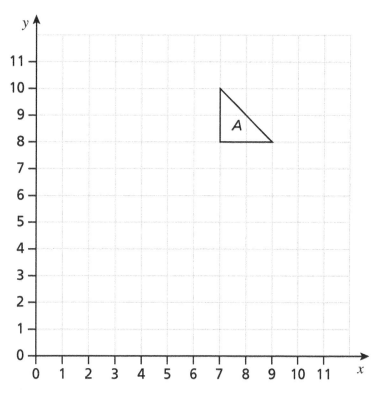

Describe the **single** transformation that will map triangle *C* to triangle *A*.

[4 marks]

_____

_____

Answer _____

20 Work out the length x in the triangle.

[3 marks]

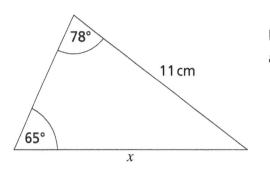

Not drawn accurately

$x =$ _____ cm

21 The diagram shows two similar bottles.

The diameter of the smaller bottle is 3 cm.

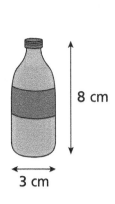

8 cm

3 cm

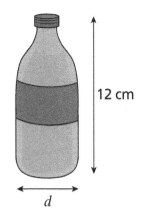

12 cm

d

21 **(a)** Work out the length d.

[3 marks]

Answer _____ cm

21 **(b)** Work out the ratio of the volumes of the smaller bottle to the larger bottle.

[2 marks]

Answer _____ : _____

22 A pyramid has a rectangular base *ABCD*.

The vertex is directly over the midpoint, *X*, of the base.

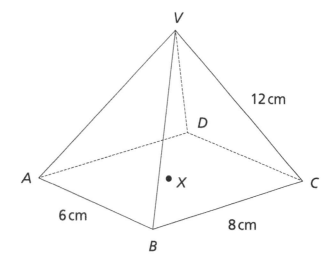

Calculate the size of the angle between the side *VC* and the base *ABCD*.

[4 marks]

Answer _____ degrees

23 **(a)** Rearrange the equation $b^3 - 2a + 3 = 0$ to make b the subject.

[2 marks]

Answer

23 **(b)** One solution of the equation $x^3 - 2x + 3 = 0$ can be found with the iterative formula

$$x_{n+1} = \sqrt[3]{2x_n - 3}$$

Starting with $x_0 = 1$, write down the value of x_1

[1 mark]

Answer

23 **(c)** Continue the iteration to find the solution.
Give your answer to 2 decimal places.

[2 marks]

Answer

24 A circle and a line are shown on the centimetre grid.

The line intersects the circle at *A*.

The circle intersects the *x*-axis at *B*.

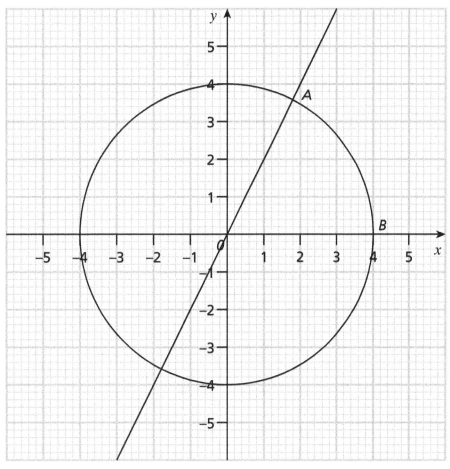

24 **(a)** Write down the equation of the circle.

[1 mark]

Answer _____

24 **(b)** Work out the length of the minor arc *AB*.

[3 marks]

Answer _____ cm

25 There are x beads in a jar.

The probability of taking a red bead from the jar at random is $\frac{4}{9}$

7 more red beads are added to the jar.

The probability of taking a red bead from the jar at random is now $\frac{1}{2}$

Use algebra to work out the value of x.

[5 marks]

Answer ....

26 Two functions are $f(x) = 3x - 1$ and $g(x) = x^2 + 2$

26 **(a)** Work out $f^{-1}(x)$

[2 marks]

Answer ....

26 **(b)** Work out $fg(x)$

[2 marks]

Answer ....

END OF QUESTIONS

Answers

Key to abbreviations used within the answers

M   method mark (e.g. M1 means 1 mark for method)

A   accuracy mark (e.g. A1 means 1 mark for accuracy)

B   independent marks that do not require method to be shown (e.g. B2 means 2 independent marks)

oe   or equivalent

ft   follow through

dep   dependent on previous mark

sc   special case

indep   independent

Question	Answer	Mark	Comments
1	$\frac{1}{3}\pi \times 6^2 \times 8$ or $\frac{1}{3}\pi \times 36 \times 8$	M1	
	96π	A1	
2	$8x^4 + 4x^3y$	B2	
3	$5x < -2 - 3$ or $5x < -5$	M1	
	$x < -1$	A1	
4	$58\frac{3}{4} - 37\frac{2}{5}$	B1	Allow one error in the numerators
	$\frac{15}{20} - \frac{8}{20}$ or $58\frac{15}{20} - 37\frac{2}{5}$	M1	Allow one error in the numerators
	$21\frac{7}{20}$	A1	
5	2×27 or 3×18	M1	Could be in a factor tree
	$2 \times 3 \times 3 \times 3$	M1dep	
	2×3^3	A1	
6	$40 \div 3$ seen or 13	M1	
	$1.25 \times$ '13' [= 16.25]	M1	Must attempt a partition method
	'16.25' + 0.48	M1	
	£16.73	A1	sc1 for £17.50
7	$a = 8$	B1	
	$b = 8 + 3 + 3 = 14$	B1	
	$c = 8 - 3 - 3 = 2$	B1	
	$d = 8$	B1	
8	$120 \div 5 \times 2$ (= 48)	M1	
	'120' – '48' (= 72)	M1dep	
	'72' ÷ [4 + 5] (= 8)	M1dep	
	40 mins	A1	

Question	Answer	Mark	Comments
9	$ADE = 58°$ or $DGC = 77°$	M1	May be labelled on the diagram
	$x = 77$ degrees	A1	
	Corresponding angles are equal **and** angles on a straight line add up to 180° Or Vertically opposite angles are equal **with** angles in a triangle add up to 180° **and** alternate angles are equal	B1	Allow two out of three reasons for B1
10 (a)	$50 \times (60 \div 6)$ (= 500) $50 \div 6 \approx 8$	M1	
	$500 \times 8 = 4000$ or $500 \times 20 = 10\,000$ $500 \times 24 = 12\,000$ '8'$\times 60 \times 8 =$ $480 \times 8 \approx 500 \times 8 = 4000$ or '8'$\times 60 \times 24$ $= 480 \times 24$ $\approx 500 \times 20$ $= 10\,000$ or '8'$\times 60 \times 24$ $= 480 \times 24$ $\approx 500 \times 25$ $= 12\,500$	A1	This answer will depend on assumptions made
10 (b)	An assumption which supports their method in part (a) e.g. 'the machine operates for 8 hours per day' or 'the machine operates for 24 hours a day'	B1	

Question	Answer	Mark	Comments
11 (a)	$4^2 + 2 \times 3 \times -2$	M1	
	$\sqrt{16 - 12}$	M1	
	$v = 2$	A1	Allow $v = 2$ and $v = -2$
11 (b)	$v^2 - u^2 = 2as$	M1	
	$a = \dfrac{v^2 - u^2}{2s}$	A1	
12	$a = -2$	B1	
	$b = 0.5$	B1	
13	2^2 or $\sqrt[3]{64}$	M1	oe
	4	A1	
14 (a) Alt 1	$\dfrac{3}{25} > \dfrac{1}{10} > \dfrac{4}{50}$	B1	
	Bob	B1dep	
14 (a) Alt 2	'Because they each did different numbers of trials'	B1	Accept similar statement
	'I can't tell'	B1dep	Accept similar statement
14 (b)	Josh	B1	
	He did the most trials	B1	Accept similar statement
15	$2n^2$	B1	
	$2n^2 - 3$	A1	
16	$4x$ or $x + 12$ seen	M1	Accept other letters used instead of 'x'
	$x + 4x = x + 12$	M1	
	3	A1	Trial and error scores zero unless final answer is correct
17 (a)	7 (6) 13 (4) (9) (13) 11 15 (26)	B3	B2 for at least one correct row and one correct column. B1 for least one correct row or one correct column.
17 (b)	$\dfrac{9}{13}$	B1ft	

Question	Answer	Mark	Comments
18 (a)	30 seconds	B1	
18 (b)	$\dfrac{1200}{5}$ or $\dfrac{1.2}{5}$	M1	
	4 m/s	A1	
18 (c)	$\dfrac{1200}{6} \div 60$ (= 3.33… mins)	M1	
	3 mins 20 seconds	B1	
	Straight line drawn from (6, 1.2) to a point marked on the x axis between 9 and 9.5	A1	Point must be > 9
19	e.g. $(x + 2)(x + 3)$ $(x - 1)$ $= (x^2 + 3x + 2x + 6)$ $(x - 1)$	M1	
	$= (x^2 + 5x + 6)$ $(x - 1)$	M1	
	$= x^3 - x^2 + 5x^2 - 5x + 6x - 6$	M1	
	$x^3 + 4x^2 + x - 6$	A1	
20 (a)	3, 8, 15, 33, 50, 57, 60	B1	Fully correct cumulative frequencies
	At least 6 points plotted from (1, 3), (2, 8), (3, 15), (4, 33), (5, 50), (6, 57), (7, 60)	B1ft	Allow follow through from part (a)
	Points joined with a smooth curve	A1	Fully correct graph
20 (b)	3.8 to 3.95 mins	B1	
20 (c)	Whisker starts at zero, LQ at 3, median at '3.8', UQ at 4.6, whisker ends at 7	B1	Allow three correct, two of which must be median and upper or lower quartile
	Fully correct box plot [ft values from their **cumulative** graph]	B1	

Question	Answer	Mark	Comments
21	Enlargement	B1	
	Scale factor −1.5	B1	
	centre (0, 3)	B1	
22	300×2^3 (= 2400)	M1	oe
	$2400 \div 1000$	M1 indep	Correct method seen to change any amount of ml into litres
	2.4 litres	A1	
23 (a)	$\dfrac{10}{7}$	B1	
23 (b)	$-\dfrac{7}{10}$	B1ft	
23 (c)	$y - 10 = -\dfrac{7}{10}(x - 7)$ or $10 = -\dfrac{7}{10} \times 7 + c$ or $c = 149$	M1	
	$10y = 149 - 7x$ or $y = -\dfrac{7}{10}x + 14.9$	A1	oe
24	$\sqrt{12} = \sqrt{3} \times \sqrt{4}$ or $\sqrt{27} = \sqrt{3} \times \sqrt{9}$	M1	
	$4\sqrt{3} + 2$	A1	

SET A – Paper 2

Question	Answer	Mark	Comments
1	350 ml = 0.35 litres or 1.2 litres = 1200 ml or 350 : 1200	M1	oe
	7 : 24	A1	
2	$\pi \times 4 \times 4$ or $\dfrac{1}{2} \times \pi \times 4 \times 4$	M1	oe
	8π	A1	
3	$\dfrac{x^4}{x^2}$ or $x^3 \times x^{-1}$	M1	
	x^2	A1	

Question	Answer	Mark	Comments
4 (a)	Primary **and** continuous	B1	With no other boxes ticked
4 (b)	Ensure each student is equally likely to be picked e.g. names in a hat	B1	Either a statement or example is acceptable
5	$\dfrac{(2x+8)(x-2)}{2}$ or $2x^2 + 8x - 4x - 16$	M1	Allow one error in the expansion
	Complete the proof to get $x^2 + 2x - 8$	A1	
6	42, 84, 126, … and 70, 140, 210, …	M1	Allow errors if intention is clear
	210 identified	M1	Or a multiple of 210
	$x = 5$ and $y = 3$	A1	Or multiples of 5 and 3
7	Any translation	B1	The shape should be exactly the same size and orientation
	Fully correct translation Top right corner should be the point (4, 4)	B1	
8 (a)	$202\,000 \times 1.015^n$ seen	M1	n can be any positive integer
	5 years	A1	
8 (b)	$180\,000 \div 1.18$ Or $180\,000 \div 1.06$	M1	
	$180\,000 \div 1.18 \div 1.06$ (= 143907)	M1	
	£144 000	A1	

Question	Answer	Mark	Comments
9 (a)	$2.176 \times 10^4 \div$ 3.2×10^7	M1	
	6.8×10^{-4}	A1	
	0.00068	B1	
9 (b)	$\left(\dfrac{1.15 \times 10^{-3}}{2.3 \times 10^{-5}} \right) \div 8$	M1	Allow two out of three terms correct
	6.25 N/m²	A1	
10	$161 \times 20 \ (= 3220)$	M1	
	$145 \times 3 + 155 \times 6$ $+165 \times 6 + 175 \times 4$ $(= 3055)$	M1	
	'3220' − '3055' (= 165)	M1dep	
	$160 < h \leqslant 170$ should have freq = 7	A1dep	Zero marks with no working
11	$(x+5)(x-3)$	M1	
	$x = 3$ and -5	A1	
12 (a)	$y = \dfrac{5x}{3} + 1$	M1	
	<table><tr><td>x</td><td>−3</td><td>0</td><td>3</td></tr><tr><td>y</td><td>−4</td><td>1</td><td>6</td></tr></table>	M1	At least one of these points correctly plotted
	Fully correct line plotted	B1	
	$x = 1.5, y = 3.5$	A1	scB1 if correct answer with no graph drawn
12 (b)	$y = -x + c$	M1	Allow gradient $= -1$
	$x + y = 7$	A1	oe
13 (a)	$0.23 \times 0.23 \times 0.77$	M1	
	0.040733	A1	Allow rounding to 0.04
13 (b)	$\sqrt{0.0961} \ (= 0.31)$	M1	
	0.69	A1	
14	$\dfrac{30}{360} \times \pi r^2 \ (= 2.5\pi)$	M1	oe
	$\sqrt{12 \times 2.5}$	M1	oe
	5.48 cm	A1	

Question	Answer	Mark	Comments
15 (a)	12 to 12:30 am	B1	
15 (b)	Tangent drawn on the graph at 10:30 pm	M1	
	Answer in range 2.4 – 2.8 (cm/h)	A1	
16 (a)	$y = k\sqrt[3]{x}$	M1	Allow $k = 2.5$ for M1
	$y = 2.5\sqrt[3]{x}$	A1	oe
16 (b)	$15 = 2.5\sqrt[3]{x}$	M1ft	
	$x = 216$	A1	
17	$455 \div 5 \ (= 91)$ **and** either 13 or 7 identified as a factor of 91	M1	
	20	A1	Allow 92 for full marks
18 (a)	$\dfrac{5a}{2} + 45a = 400$	M1	oe
	8.42 m/s	A1	
18 (b)	1.5×10 or 0.4×5 or 0.4×15	M1	
	$1.5 \times 10 + 0.4 \times 5$ $+ 0.4 \times 15$	M1	
	23	A1	
19 (a)	$4\left[x^2 - \dfrac{5}{4}x + 3 \right]$	M1	
	$4\left[\left(x - \dfrac{5}{8} \right)^2 - \dfrac{25}{64} + 3 \right]$	M1	
	$\left(\dfrac{5}{8}, 10\dfrac{7}{16} \right)$ oe	A2	1 mark for each
19 (b)	(5, 3)	B2	1 mark for each
20	UB = 50.005 m, LB = 49.995 m UB = 135.5 s, LB = 134.5 s	M1	At least one correct
	$\dfrac{200.02}{134.5}$ or $\dfrac{199.98}{135.5}$	M1dep	oe
	1.487(137546) or 1.475(867159)	B1dep	
	1.5 m/s	A1dep	No marks if 1.5 comes from $\dfrac{4 \times 50}{135}$

Question	Answer	Mark	Comments
21	$2n(2n + 2)(2n + 4)$	M1	At least two correct expressions for even, consecutive numbers
	$8n^3 + 16n^2 + 8n^2 + 16n$	M1ft	At least two terms correct
	$8(n^3 + 3n^2 + 2n)$	A1	
22	$\overrightarrow{BC} = \dfrac{3}{4}\mathbf{b}$	M1	
	$\overrightarrow{CE} = \dfrac{1}{8}\mathbf{b}$	M1	
	$\overrightarrow{AE} = \overrightarrow{AB} + \overrightarrow{BC} + \overrightarrow{CE}$	M1	
	$\overrightarrow{AE} = \mathbf{a} + \dfrac{7}{8}\mathbf{b}$	A1	oe
23	$CD = \dfrac{10.8\sin 65}{\sin 61}$ $(= 11.191...)$	M1	
	$\sin\widehat{CBD} = \dfrac{'CD' \times \sin 54}{9.1}$ $(= 0.994...)$	M1dep	
	$\widehat{CBD} = \sin^{-1}\left(\dfrac{'CD' \times \sin 54}{9.1}\right)$ $(= 84.233...)$	M1dep	
	$\dfrac{1}{2} \times 9.1 \times 'CD' \times \sin'41.766...'$	M1dep	
	$33.9\ \text{cm}^2$	A1	

Question	Answer	Mark	Comments
1	£350 × 0.87	M1	
	£304.50	A1	
2	Geometric	B1	
3	$2x + 8 = y$	M1	
	Gradient = 2	A1	
4	180 + 60	M1	
	240°	A1	
5	Scale factor = 1.5 or $\dfrac{2}{3}$ or $\dfrac{8}{5}$ or $\dfrac{5}{8}$ or 8 × 1.5 or $8 \div \dfrac{2}{3}$ or $7.5 \times \dfrac{8}{5}$ or $7.5 \div \dfrac{5}{8}$	M1	
	12 cm	A1	
6 (a)	Fully correct	B2	B1 for at least six points plotted correctly
6 (b)	No correlation	B1	
	No connection between height and salary	B1	
7	Attempt at a method to find prime factors for both $135 = 3 \times 3 \times 3 \times 5$ $630 = 2 \times 3 \times 3 \times 5 \times 7$	M1	Accept at least one correct step for each
	Either $3 \times 3 \times 3 \times 5$ or $2 \times 3 \times 3 \times 5 \times 7$ or $3 \times 3 \times 5$ seen	M1 indep	At least one fully complete
	HCF = 45	A1	
8	51, 30, 13, 17, 21, 15, 6	M1	1 mark for at least three correct entries
	Fully correct diagram	A1	2 marks for fully correct
9	$\dfrac{26 - 19}{26} \times 100$	M1	
	26.9%	A1	Allow 27%

Question	Answer	Mark	Comments
10	Complete method seen e.g. $\dfrac{19}{5} \times \dfrac{4}{3}$	M1	oe
	$\dfrac{76}{15}$	A1	
	$5\dfrac{1}{15}$ inches	B1	
11	(exterior angle =) $180 - 2x$	M1	
	$\dfrac{360}{180 - 2x}$	M1	
	$\dfrac{180}{90 - x}$	A1	
12	2×10.85 (= 21.70)	M1	A4 print is free
	$21.70 - (2 \times 3.09 + 1.52 + 3.80)$ [= 10.20]	M1	Allow 30.35 in place of 21.70
	$2 \times 5.95 - (2 \times 1.07 + 3.80)$ [= 5.96]	M1	
	$8.65 - (1.52 + 2.40)$ [= 4.73]	M1	
	'10.20' + '5.96' + '4.73'	M1dep	
	£20.89	A1	
13	Perpendicular bisector of Brooks and Redding constructed	M1	Arcs should be visible
	Arc / Circle about Dufresne with radius of 3.1 cm	M1	Accept 3 → 3.2 cm
	Correct region shaded bounded by 'arc' and 'bisector'	A1dep	Dependent on at least one M1
14	$2x^2 - 5x - 3 \leqslant 0$	M1	Allow '=' in place of '⩽'
	$(2x + 1)(x - 3)$	M1dep	
	−0.5 or 3 identified as boundary solutions	A1dep	
	$-0.5 \leqslant x \leqslant 3$	A1	

Question	Answer	Mark	Comments
15	5, 6, 13, 15, 15, 17, 17, 18, 20, 21, 26	M1	Ordering the numbers
	Median = 17	A1	
	Upper quartile = 20 Lower quartile = 13	A1	
	Yes, with 13, 17 and 20, or yes, differences are 4 and 3	A1	
16	2500 ml or 0.5 litres seen	M1	
	$2500 \times (0.965)^n$	M1	Any positive value of n tried
	7 minutes	A1	
17 (a)	Bottom right diagram circled	B1	
17 (b)	A (parabolic) curve starting at zero and getting steeper	B1	
18	Angle BOC = 76° and Angle at centre is twice the angle at the circumference. or Angle BAC = 38° and Angles (subtended) on same arc are equal.	B2	B1 for angle BOC = 76° or angle BAC = 38°
	x = 104° and Angles on straight line add up to 180°. or x = 104° and Angles in triangle (ABO) add up to 180°.	B2	B1 for x = 104°

Question	Answer	Mark	Comments
19	$x = 0.2333...$ or $10x = 2.333...$ or $100x = 23.333...$	M1	
	$90x = 21$	M1dep	
	$\dfrac{21}{90} = \dfrac{7}{30}$	A1dep	
20 (a)	$5x = 3 - x^3$	M1	Attempt to add 3 and subtract $5x$ from both sides
	$x = \dfrac{3 - x^3}{5}$	A1	
20 (b)	$0^3 + 5 \times 0 - 3 = -3$ AND $1^3 + 5 \times 1 - 3 = 3$	M1	
	Sign changes, therefore x must lie between 1 and 0	B1	oe
20 (c)	$x_1 = \dfrac{3 - 0}{5}$ $(= 0.6)$	M1	
	$x_2 = \dfrac{3 - '0.6^3\,'}{5}$ $(= 0.556...)$	M1dep	
	$x_3 = 0.565...$, $x_4 = 0.563...$ and $x_5 = 0.564...$ with 0.56 identified as the final answer to 2 decimal places	A1dep	
21 (a)	Even only : 4, 10, 20, 50, 100 Prime only: 5	B1	
	Intersection: 2	B1	
	Outside the circles: 1 and 25	B1	
21 (b)	$\dfrac{1}{9}$	A1	

Question	Answer	Mark	Comments
22	$\left(\sqrt{2}\right)^n$ or $\left(\sqrt{2}\right)^9$ seen	M1	
	$16\sqrt{2}$	A1	
23	$x + 1 = x^2 - 5x + 10$	M1	
	$x^2 - 6x + 9 = 0$	M1	
	$(x - 3)(x - 3) = 0$	M1	
	$x = 3$, $y = 4$	A1	
24	$5(x^2 - 4) = 5(x - 2)(x + 2)$	M1	
	$x(x - 2)$	M1	
	$\dfrac{5(x + 2)}{x}$ or $\dfrac{5x + 10}{x}$	A1	
25 (a)	$AG = \sqrt{1^2 + 1^2}$ $(= \sqrt{2})$	M1	'1' could be replaced by any other chosen value for the side length of the cube
	$AF = \sqrt{\left(\sqrt{2}\right)^2 + 1^2}$ $(= \sqrt{3})$	M1 depft	ft from their chosen value for '1'
	$1 : \sqrt{3}$	A1	
25 (b)	$\tan^{-1}\left(\dfrac{1}{'\sqrt{2}\,'}\right)$	M1ft	Or their values for '1' and '$\sqrt{2}$' in part (a)
	35.3 degrees	A1	

SET B – Paper 1

Question	Answer	Mark	Comments
1 (a)	-3.5	B1	
1 (b)	$f^{-1}(x) = x + 3$	B1	
2	$x = 2$ and $x = -3$	B1	
3 (a)	$6^2 + 4^2$ or $36 + 16$ or 52	M1	
	$\sqrt{52}$ or $2\sqrt{13}$	A1	
3 (b)	$\dfrac{3}{4}$	B1	
4	$6x - 12 + 8 = x$	M1	
	$5x = 4$	M1dep	
	$x = 0.8$	A1	oe
5	Area of any face, i.e. 20×5 or 100 etc.	M1	
	$2 \times 100 + 2 \times 50 + 2 \times 200$	M1dep	
	$700\,\text{cm}^2$	A1	
6	$4x + 4 - 6x + 8$	M1	M1 for three terms correct
	$4x + 4 - 6x + 8$	A1	A1 for four terms correct
	$-2x + 12$	A1ft	
7 (a)	$230\,000$	B1	
7 (b)	5×10^{-4}	B1	
7 (c)	1.6×10^8	B2	B1 for 16×10^7
8 (a)	-1.5 and 3	B2	B1 each answer
8 (b)	$(0.75, -6.1)$	B1	
9	$2x + 100 = 180$	M1	
	$360 \div 40$	M1dep	
	9	A1	
10	$x + 2 = 2x - 1$	M1	
	$x = 3$	A1	
	Side $= 5\,\text{cm}$	B1	
	Area $= 25\,\text{cm}^2$	A1	
11	$x^2 + 2x + 1$ or $x^2 - 2x - 3$	M1	
	$x^3 - 3x^2 + 2x^2 - 6x + x - 3$	M1dep	
	$x^3 - x^2 - 5x - 3$	A1	
12	$\pi \times x^2 \times 3 = 48\pi$	M1	
	$3x^2 = 48$ or $x^2 = 16$	M1	
	$x = 4\,\text{cm}$	A1	

Question	Answer	Mark	Comments
13	$x \leqslant 6$	B1	
	$x + y \geqslant 7$	B1	
	$y \leqslant x + 1$	B1	
14	$27 + 9\sqrt{2} - 3\sqrt{8} - \sqrt{16}$	M1	oe
	$27 + 9\sqrt{2} - 6\sqrt{2} - 4$	A1	
	$23 + 3\sqrt{2}$	A1	
15	Vertical scale marked to at least 3.5 Bar between 5–10 to a height of 3 Bar between 10–20 to a height of 3.5 Bar between 20–35 to a height of 2 Bar between 35–45 to a height of 1.5 Bar between 45–50 to a height of 1	B3	B2 Scale marked and any two bars B1 Scale marked and any 1 bar
16 (a)	56 degrees	B1	
16 (b)	ACB stated or shown as 32	B1	
	CAB stated or shown as 90 (may be implied by working)	B1	
	58 degrees	B1	
17	16	B2	B1 for $(\sqrt[3]{64})^2$ oe B1 for $\sqrt[3]{64} = 4$
18 (a)	24	B1	
18 (b)	31 and 17 seen	M1	
	14	A1	
18 (c)	Valid box plot with Median marked (ft their median) IQR marked (ft their IQR) Minimum value as 5 and maximum as 50	B2	B1 any two components

Question	Answer	Mark	Comments
19 (a)	$\mathbf{a} + \dfrac{3}{2}\mathbf{b}$	B1	
19 (b)	$\overrightarrow{BC} = \overrightarrow{BA} + \overrightarrow{AO}$ $+ \overrightarrow{OC} = -\mathbf{a}$ $+ \dfrac{1}{2}\mathbf{b}$ or $-\dfrac{3}{2}\mathbf{b} - \mathbf{a} + OC$ $= -\mathbf{a} + \dfrac{1}{2}\mathbf{b}$	M1	
	$2\mathbf{b}$	A1	
19 (c)	Yes, OC is a multiple of AB so parallel.	B1	oe
20	$x = 0.733333...$ and $10x$ $= 7.33333$	M1	
	$9x = 6.6$ or $\dfrac{66}{90}$	A1	
	$3\dfrac{11}{15}$	A1	
21	$\dfrac{x^2}{2} = 9$	M1	
	$x = 3\sqrt{2}$	M1	
	Hypotenuse $= 6$	M1	
	$6 + 2 \times 3\sqrt{2}$	M1	
	$6 + 6\sqrt{2}$ cm	M1	

Question	Answer	Mark	Comments
22	Tree diagram with at least 3 correct probabilities marked or P(R and B) + P (B and R)	M2	M1 for less than three correct probabilities marked
	All correct probabilities identified as $\dfrac{7}{10}$, $\dfrac{3}{10}$, $\dfrac{6}{9}$ oe, $\dfrac{3}{9}$ oe, $\dfrac{7}{9}$ and $\dfrac{2}{9}$ or one of $\dfrac{7}{10}$ $\times \dfrac{3}{9}$ or $\dfrac{3}{10} \times \dfrac{7}{9}$	A1	
	$\dfrac{7}{10} \times \dfrac{3}{9} + \dfrac{3}{10}$ $\times \dfrac{7}{9}$	M1dep	
	$\dfrac{42}{90}$ or $\dfrac{7}{15}$	A1	
23	$(2x + 3)(2x - 5)$	M1	Allow $(2x + a)(2x + b)$ where $ab = -15$
	$(2x + 3)(2x - 3)$	M1	Allow $(2x + a)(2x + b)$ where $ab = -9$
	$\dfrac{2x - 5}{2x - 3}$	A1	
24	Gradient AB $= -\dfrac{1}{2}$	M1	
	Gradient perpendicular 2	A1	
	Midpoint AB $= (5, 9)$	B1	
	$9 = 2 \times 5 + c$ or $c = -1$ or $y - 9 = 2(x - 5)$	M1	
	$y = 2x - 1$	A1	

SET B – Paper 2

Question	Answer	Mark	Comments
1	9	B1	
2	Any fraction that gives a recurring decimal, e.g. $\dfrac{1}{3}$	B1	Note, the denominators of the first few unit fractions having repeating decimals are 3, 6, 7, 9, 11 and 12.
3	(–6, 7)	B1	
4 (a)	135 degrees and alternate angles (are equal)	B1	
4 (b)	114 degrees and corresponding angles (are equal)	B1	
5	Correct translation, i.e. $(1, 1) \rightarrow (-2, -3)$, etc.	B2	B1 for correct translation of one vector component
6	$6^2 + 11^2$ or 36 + 121	M1	
	$\sqrt{6^2 + 11^2}$ or $\sqrt{36 + 121}$	M1dep	
	$\sqrt{157}$ or 12.5... cm	A1	
7	5 × 145 + 9 × 155 + 12 × 165 + 8 × 175 + 6 × 185 or 6610	M1	
	6610 ÷ 40	M1dep	
	165.25 cm	A1	
8 (a)	Any product including a prime that makes 28	M1	
	2 × 2 × 7 or $2^2 \times 7$	A1	
8 (b)	2 × 2 × 5 × 7	M1	
	140	A1	
9	$4(x + 4) = 26$	B1	
	$4x + 16 = 26$ or $4x = 26 - 16$	M1	
	$4x = 10$	M1dep	
	$x = 2.5$	A1	
10	0.85	B1	
	238 ÷ 0.85	M1	
	£280	A1	

Question	Answer	Mark	Comments
11	36 ÷ 3 or 12	M1	
	2 × 12 or 5 × 12	M1dep	
	24 and 60	A1	
12	$\sqrt{\dfrac{402}{\pi}}$ or 11.3...	M1	
	$11.3 \times \pi + 2 \times 11.3$	M1dep	
	[58, 58.2] cm	A1	
13	Arc from A cutting given line	M1	
	Arc centred on intersection and crossing original arc plus line drawn and angle 60° drawn	A1	
	60° angle bisected	A1	Angle must be between [28, 32]
14 (a)	$5(4x^2 + 3x - 8x - 6)$	M1	
	$5(4x^2 - 5x - 6)$ or $20x^2 + 15x - 40x - 30$	M1dep	
	$20x^2 - 25x - 30$	A1	
14 (b)	$2(x + a)(x + b)$	M1	$ab = \pm3$
	$2(x + 1)(x + 3)$	A1	oe e.g. $(2x + 2)(x + 3)$
15	Triangle between (3, 9), (4, 9) and (4, 7)	B3	B2 two vertices correct B1 rays marked through (5, 8)
16	30 × 1.6 or 48	M1	
	(their 48 – 40) ÷ 40 (× 100)	M1dep	
	20	A1	
17 (a)	$(x + 3)^2$	M1	
	$(x + 3)^2 - 9$	M1dep	
	$(x + 3)^2 - 18$	A1	
17 (b)	$x + 3 = \sqrt{18}$	M1	
	$x = -3 \pm \sqrt{18}$	A1	
18	$2(4x - 1) - 3(x + 1)$	M1	
	$5x - 5 =$	A1	
	$(4x - 1)(x + 1)$ or $4x^2 + 4x - x - 1$	M1	
	$4x^2 - 2x + 4$	A1	

Question	Answer	Mark	Comments
19 (a)	$y = kx^2$ and $20 = k \times 2^2$	M1	
	$k = 5$	A1	
	500	A1	
19 (b)	$5 = 5 \times x^2$	M1	
	± 1	A1	Condone omission of $\pm$
20	$x(x - 6) + 2x + x - 4 + x + 30 = 146$	M1	
	$x^2 - 2x - 120 = 0$	A1	
	$(x - 12)(x + 10) = 0$	A1	
	$x = 12$	A1	
	$\dfrac{8}{146}$ or $\dfrac{4}{73}$	A1	
21 (a)	$\cos x = \dfrac{10^2 + 7^2 - 13^2}{2 \times 10 \times 7}$	M1	
	$-\dfrac{1}{7}$	A1	
	98.2 degrees	A1	
21 (b)	$\dfrac{1}{2} \times 7 \times 10 \times \sin$ (their 98.2)	M1	
	34.6... cm²	A1	
22 (a)	Tangent drawn at 3	M1	
	y-step and x-step measured	M1dep	
	[0.7, 1.1] m/s²	A1ft	ft their tangent
22 (b)	Attempt to calculate area under curve	M1	
	[75, 85]	A1ft	ft their area
	Their area ÷ 10	M1dep	
	[7.5, 8.5] m/s	A1	
23	11.5 or 12.5 or 8.25 or 8.35	M1	
	11.5 and 12.5 and 8.25 and 8.35	M1dep	
	8.25 ÷ 12.5 or 8.35 ÷ 11.5	M1	
	Upper 46.6	A1	
	Lower 41.3	A1	

SET B – Paper 3

Question	Answer	Mark	Comments
1	125	B1	
2	1 : 1.8	B1	
3	343	B1	
4	56 ÷ 8	M1	
	7	A1	
5 (a)	x^9	B1	
5 (b)	x^{10}	B1	
6	$\begin{pmatrix} 10 \\ 4 \end{pmatrix}$	B2	B1 for each component
7	–2, –1, 0, 1, 2, 3	B2	B1 for –2, –1, 0, 1, 2, 3, 4 or –1, 0, 1, 2, 3
8		B2	B1 for any enlargement that reduces the size of the shape and keeps the side in relative ratio. B1 for any three sides correct.
9	$3x + 2y = 2$ and $3x + 12y = 27$ or $6x + 4y = 4$ and $x + 4y = 9$	M1	
	$x = -1$	A1	
	$y = 2.5$	A1	
10 (a)	$\dfrac{4}{10}$ marked on red and $\dfrac{6}{10}$ marked on blue	B1	
10 (b)	$\dfrac{4}{10} \times \dfrac{4}{10}$ or $\dfrac{6}{10} \times \dfrac{6}{10}$	M1	
	$\dfrac{4}{10} \times \dfrac{4}{10} + \dfrac{6}{10} \times \dfrac{6}{10}$	M1dep	
	0.52	A1	oe
11	1.5 ÷ 2	M1	
	0.75 Pa/hour	A1	

Question	Answer	Mark	Comments
12 (a)	$(x + 5)(x - 5)$	B1	
12 (b)	$x^2 + 4x + 4$ or $x^2 + 2x + 1$	M1	$(x + 2 + x + 1)$ $(x + 2 - (x + 1))$
	$x^2 + 4x + 4 - (x^2 + 2x + 1)$	M1dep	$(2x + 3)(1)$
	Shows subtraction of terms clearly	A1	
13 (a)	$\sin 32° = \dfrac{x}{12}$ or $12 \times \sin 32°$	M1	
	$12 \times \sin 32°$ $= 6.359...$ $= 6.36$ (2 d.p.)	A1	
13 (b)	$\pi \times 6.36 \times 12$	M1	
	$[236.6, 240]\,\text{cm}^2$	A1	
14	0.92 seen	B1	
	Any value of 0.92^n or 6000×0.92^n calculated where $n > 3$	M1	
	$0.92^8 = 0.51...$ and $0.92^9 = 0.47...$ or $6000 \times 0.92^8 = 3079$ And $6000 \times 0.92^9 = 2832$	A1	
	9 years	A1	Accept just over 8 or between 8 and 9
15	$A = \tan x$ $B = \dfrac{1}{x}$ $C = 2^x$	B2	B1 for one correct
16	$4x^4y^6$	B2	B1 for two parts correct
17 (a)	B and D	B1	
17 (b)	A and D	B1	
18 (a)	30 and 38	B2	B1 each
18 (b)	Works out second difference as 2	M1	
	Subtracts n^2 from series, i.e. 5, 6, 7, 8, 9 ...	M1dep	
	Identifies $n + 4$ as linear sequence	A1	
	$n^2 + n + 4$	A1	

Question	Answer	Mark	Comments
19	Shows reflected triangle B at (7, 2), (7, 4) and (9, 4)	M1	
	Shows reflected triangle C at (1, 4), (3, 4) and (3, 2)	M1dep	
	Rotation, 180°, about (5, 6)	A2	A1 for two parts. Accept reflection in line $y + x = 11$ oe
20	$\dfrac{x}{\sin 78} = \dfrac{11}{\sin 65}$	M1	
	$x = \dfrac{11 \times \sin 78}{\sin 65}$	M1dep	
	$[11.87, 11.9]\,\text{cm}$	A1	
21 (a)	Scale factor is 1.5 or $\dfrac{2}{3}$ or $\dfrac{8}{3}$ or $\dfrac{3}{8}$	M1	
	3×1.5 or $3 \div \dfrac{2}{3}$ or $12 \div \dfrac{8}{3}$ or $12 \times \dfrac{3}{8}$	M1dep	
	4.5 cm	A1	
21 (b)	$2^3 : 3^3$	M1	oe
	8 : 27	A1	
22	$AC^2 = 6^2 + 8^2$ or $CX^2 = 3^2 + 4^2$	M1	
	$AC = \sqrt{6^2 + 8^2} = 10$ or $CX = \sqrt{3^2 + 4^2} = 5$	A1	
	$\cos VCX = \dfrac{5}{12}$	M1	
	Angle $VCX =$ 65.37... degrees or 65.4 degrees or 65 degrees	A1	
23 (a)	$b^3 = 2a - 3$	M1	
	$b = \sqrt[3]{2a - 3}$	B1	
23 (b)	−1	B1	
23 (c)	−1.89	B2	B1 for any further iterations or 1.89....

Question	Answer	Mark	Comments
24 (a)	$x^2 + y^2 = 16$	B1	
24 (b)	Angle = $\tan^{-1}(2)$ or $63.43..$	M1	
	(their $63.43 \div 360$) $\times 2 \times \pi$ $\times$ their radius	M1dep	
	[4.36, 4.43] cm	A1	
25	$\dfrac{4}{9}x$	M1	
	$\dfrac{4}{9}x + 7$	M1dep	
	$\dfrac{4}{9}x + 7 = \dfrac{x+7}{2}$	M1dep	
	$\dfrac{1}{18}x = \dfrac{7}{2}$	M1dep	
	63	A1	T&I B1 for correct answer
26 (a)	$\dfrac{x+1}{3}$	B2	B1 for numerator of $3(x + 1)$ B1 for $\dfrac{x-1}{3}$
26 (b)	$3(x^2 + 2) - 1$	M1	
	$3x^2 + 5$	A1	